Ginger, Turmeric, and Indian Arrowroot

Growing Practices and Health Benefits

Roby Jose Ciju

DEDICATION

This book is dedicated to all home gardeners, farmers, commercial growers and all those who have a genuine interest in growing plants...

CONTENTS

ACKNOWLEDGMENTS

As we know, everything worthwhile is always accomplished through team effort only. I would like to acknowledge the efforts of my staff at agrihortico.com for helping me to republish this version of my book on "Ginger, Turmeric and Arrowroot" with additional information and relevant images. Hope my readers will find this book very useful in their gardening/farming endeavors...

INTRODUCTION OF GINGER FAMILY

Zingiberaceae, the ginger family is known for its prominent members such as ginger, cardamom, turmeric, galangal, Indian arrowroot, white turmeric or zedoary, and Myoga ginger or Japanese ginger. Ginger family is a group of about 1600 identified, flowering plant species that are grouped across 50 plant genera. Almost all members of ginger family are widely distributed throughout tropical South East Asian countries, tropical Africa, and the Americas.

Plants belonging to the family Zingiberaceae are a group of low-growing, (some plants reach up to one meter in height), aromatic perennial herbaceous plants. The basal sheaths of the plant overlap to form a pseudostem, which a characteristic feature of all members of the ginger family. Another major distinguishing feature of these plants is their flowers. Flowers are hermaphroditic and zygomorphic and are borne in determinate cymose inflorescences. Large, conspicuous, green-coloured bracts are spirally arranged, enfolding flowers. The perianth or the outer part of the flower is composed of two whorls, a fused tubular calyx, and a tubular corolla.

Figure 1: Flowers of Ginger Family

Commercially important plants such as ginger and turmeric are often grown as annual herbs and their edible rhizomes are often harvested after 9-10 months of growing.

Ginger (Zingiber officinale), cardamom (Amomum elettaria) and turmeric (Curcuma longa) are often grown as food and medicinal crops. However, a large number of ginger family members are grown as ornamental plants. Major ornamental plants are Alpinia spp., Curcuma alismatifolia, Curcuma zeodaria, and Ginger Lily (Hedychium spp.)

Among plants belonging to Alpinia species, *Alpinia purpurata* or Red Ginger Plant is a highly popular ornamental plant among the gardeners and florists for its large, showy and attractive flowers which are widely used in various types of fresh flower arrangements. Red ginger plant is native to Malaysia and surrounding regions.

Figure 2: Red Ginger Plant

Another popular ornamental plant in the tropical gardens across the world is "ginger lily". It is also known as garland flower. Ginger lily belongs to the genus "Hedychium" of the Zingiberaceae family. Ginger lilies are known for its white, fragrant flowers.

Figure 3: Hedychium coronarium

Curcuma zedoaria is also known as white turmeric or zedoary. It is popular as a house plant in many tropical countries including India. Its rhizomes may be used for food purposes also. In India, its fresh rhizomes are peeled and cleaned before slicing them. The thin slices of zedoary are then used for making various types of pickles. In Indonesia and other South East Asian countries, zedoary rhizomes are dried and powdered to make zedoary powder. Zedoary powder may be used as a spice. It is an ingredient in many curry pastes also. Zedoary powder is used as a spice in many Western countries. In Thailand, fresh zedoary rhizomes are used in fresh salads for raw consumption.

There are some prominent medicinal plants as well. Major medicinal plants in the ginger family include members like

Galangal or Thai ginger (Alpinia galangal) and Myoga (Zingiber mioga).

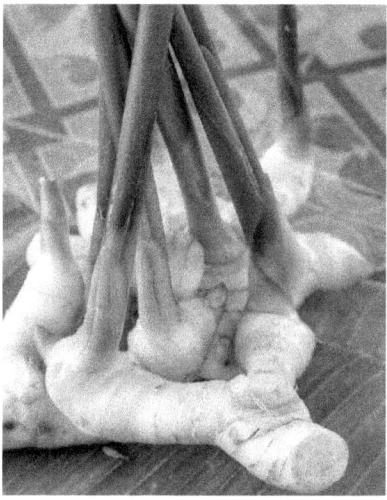

Figure 4: Galangal

Galangal is also known as "Thai Ginger" because of its extensive use in various Thai cuisines. Galangal is available in the market as fresh rhizomes, galangal powder, and dried rhizomes. Galangal powder is widely used in traditional Chinese medicines.

Myoga ginger is a traditional crop in Japan and therefore also known as "Japanese Ginger". It is very popular in China and Korea also. Myoga is believed to have anticarcinogenic properties. In Myoga ginger, tender and flavorful flower buds and shoots are used for edible purposes.

Figure 5: Flower Buds of Myoga Ginger

GINGER

Scientific name of ginger is *Zingiber officinale*. Synonyms are *Amomum zingiber* and *Zingiber missionis*. It belongs to the family Zingiberaceae. Other members of Zingiberaceae family are turmeric (Curcuma longa), cardamom (Elettaria cardamomum), white turmeric (Curcuma zedoaria), bitter ginger (Zingiber zerumbet), myoga (Zingiber mioga) and galangal, a medicinal plant.

Figure 6: A Ginger Plantation

Ginger is mainly used as a spice and as a herbal medicine. The characteristic aromatic fragrance and spicy flavor of ginger is due to the presence of volatile oils such as zingerone, shogaols and gingerols.

Ginger is actually an herbaceous perennial crop but for commercial production it is grown as an annual herb. Ginger is tropical and subtropical in its growth habit. It is grown for its aromatic rhizomes which are used as a vegetable, a spice and as a traditional medicine. Ginger rhizomes are often called 'ginger root' though it is not actually a root. Rhizome is actually a modified underground stem of the plant.

Origin and Distribution: Ginger is believed to be originated in South East Asia, particularly in the region comprising of India and China.Today, India is world's largest producer of ginger. Other major ginger-producing countries are Nigeria, China, and Indonesia.

Taxonomy: A description of taxonomic classification of ginger is given below:

Kingdom	Plantae
Class	Equisetopsida
Subclass	Magnoliidae
Superorder	Lilianae
Order	Zingiberales
Family	Zingiberaceae
Genus	Zingiber
Species	officinale

Botanical Description: Ginger is a low-growing herbaceous plant with tender stems and leaves. New plant grows from rhizomes planted underground. Upon maturity, each ginger plant reaches up to a height of one meter. Stem is covered by leaf sheaths. Leaves are light green in colour; and long and lanceolate in shape with a marked midrib. Each leaf is about 15 cm in length. Small yellowish flowers are borne in dense spikes. The most important plant part is edible rhizome. Fresh rhizome has a pale yellowish interior with skin colour varying from brownish dark to off-white.

The plants reach maturity within 8-10 months after planting. Harvesting time can be determined when the green leaves begin to turn yellow and stalks start withering. Harvesting is normally done after all leaves and stalks are dried and withered away.

Figure 7: Ginger Rhizomes/Roots: The Edible Part of the Plant

Economic Importance of Ginger: Ginger is a highly important and economically significant food and medicinal crop. Ginger is used for food and drink purposes, medicinal purposes and cosmetic uses. Raw ginger is used as a vegetable and as a food flavoring agent. Ground ginger paste is used to add taste and flavor to non-vegetarian food preparations. Dry ginger has a great commercial significance as commercial-grade ginger powder is made from dry ginger. Ginger powder is used in many traditional medicinal preparations, particularly in

Ayurvedic medicines. Ginger powder has wide applications in bakery and confectionery industry also. Ginger oil extracted from dried ginger roots are extensively used in cosmetics industry.

Ginger Products: Various ginger products available in the market are raw ginger, ginger paste, dry ginger, ginger powder, ginger oil, ginger oleoresin, ginger candy, ginger wine, bleached dry ginger, ginger beer, brined ginger, and ginger preserves.

Figure 8: Dried Ginger

Dry Ginger: Fresh ginger rhizomes are peeled and dried in the sun to obtain commercially important 'dry ginger'. Dry ginger is marketed either as 'black' or 'white' where black dry ginger is dried ginger rhizomes with its skin on while white dry ginger is dried ginger rhizomes with its skin peeled off. Dried ginger is available in the market as whole or sliced or powdered forms. Dry ginger powder is a major ingredient in many curry powder preparations and Ayurvedic medicinal preparations. In many parts of the world, powdered dried ginger or ginger powder is used to prepare ginger bread, ginger cookies, ginger cakes and ginger biscuits.

Figure 9: Ginger Powder

Food Uses of Ginger: A detailed account of various food uses of ginger and ginger products is given below:

Raw Ginger as a Vegetable and a Fiery, Pungent and Spicy Food Flavouring Agent: Raw ginger is known as 'fresh ginger' which imparts the freshest and purest ginger flavour to food preparations. Young and fresh ginger roots are fleshy and succulent with a mild flavour and aroma. Normally, young and tender ginger roots are used for food preparations as mature ginger roots tend to be dry and fibrous. Sliced ginger is used for pickle preparations, chutneys and for making ginger curry pastes. Ginger pickles and various types of ginger chutneys are highly favoured food delicacies in many South East Asian countries. Ginger paste is used for making thick gravy for spicy vegetable, egg, fish and meat preparations.

According to USDA, fresh ginger is a powerhouse of nutrients and a detailed account of nutrition in 100 grams of edible portion of fresh ginger is given below:

Nutrient	Unit	Value/100 g
Water	g	78.89
Energy	kcal	80
Protein	g	1.82
Total lipid (fat)	g	0.75
Carbohydrate	g	17.77
Fiber, total dietary	g	2
Sugars, total	g	1.7
Calcium, Ca	mg	16
Iron, Fe	mg	0.6
Magnesium, Mg	mg	43
Phosphorus, P	mg	34
Potassium, K	mg	415
Sodium, Na	mg	13
Zinc, Zn	mg	0.34
Vitamin C	mg	5
Thiamin	mg	0.025
Riboflavin	mg	0.034
Niacin	mg	0.75
Vitamin B-6	mg	0.16
Folate, DFE	µg	11
Vitamin A, IU	IU	0
Vitamin E (alpha-tocopherol)	mg	0.26
Vitamin K (phylloquinone)	µg	0.1
Fatty acids, total saturated	g	0.203
Fatty acids, total monounsaturated	g	0.154
Fatty acids, total polyunsaturated	g	0.154

Ground Ginger for Culinary Purposes: Ground ginger is used as a spice in culinary preparations. It adds flavour, taste and thick gravy to many spicy food preparations particularly non-vegetarian food preparations such as meat, egg and fish preparations. Nutrition in ground ginger is given below:

Nutrient	Unit	Value/100 g
Water	g	9.94
Energy	kcal	335
Protein	g	8.98
Total lipid (fat)	g	4.24
Carbohydrate	g	71.62
Fiber, total dietary	g	14.1
Sugars, total	g	3.39
Calcium, Ca	mg	114
Iron, Fe	mg	19.8
Magnesium, Mg	mg	214
Phosphorus, P	mg	168
Potassium, K	mg	1320
Sodium, Na	mg	27
Zinc, Zn	mg	3.64
Vitamin C	mg	0.7
Thiamin	mg	0.046
Riboflavin	mg	0.17
Niacin	mg	9.62
Vitamin B-6	mg	0.626
Folate, DFE	µg	13
Vitamin A, IU	IU	30
Vitamin E (alpha-tocopherol)	mg	0
Vitamin K (phylloquinone)	µg	0.8
Fatty acids, total saturated	g	2.599
Fatty acids, total monounsaturated	g	0.479
Fatty acids, total polyunsaturated	g	0.929

Using Fresh Ginger as Ginger Preserves and Ginger Candies: Fresh, tender ginger rhizomes are cleaned and peeled before slicing them into small pieces. These small chunks of ginger are cooked in sugar syrup. After cooling the syrup, this hot and spicy ginger preserve is bottled in air tight containers and kept for further use. Ginger chunks immersed in sugar syrup are dried/dehydrated for crystallisation in order to make ginger candies.

Ginger Oil (Gingerol) for Medicinal and Cosmetic Benefits: Gingerol, an essential oil extracted from both fresh and dried ginger rhizomes have numerous medicinal and aromatic uses. Gingerol is sudorific (encourages sweating); expectorant (loosens and expels phlegm from lungs); stomachic (improves digestion and cures stomach disorders); and Emmenagogue (brings on menstruation). Ginger oil encourages sweating in high temperature fevers and cures cough and common cold.

Beverage Uses of Ginger: There are a number of ginger-based beverages available in the market today. Some of them are ginger ale, ginger beer, ginger wine and ginger herbal tea.

How to Prepare Ginger Herbal Tea: Fresh ginger root is used to prepare ginger tea. Clean, washed and peeled ginger roots are either sliced or crushed before adding to luke warm water and then water is boiled until all ginger essence is released into the water. Finally, other ingredients such as milk, sugar and tea leaves are added. Boil this mixture once again and ginger tea is ready. This is a strong tea and in India it is called 'atrak chai' which is a specialty tea during winter months. Regular consumption of 'atrak chai' during winter months protects a person from cold-weather related sicknesses such as cough, cold, fever, sore throats, and breathing difficulties. Ginger tea has lots of other medicinal benefits too. It alleviates stomach pain and soothes the digestive system. Ginger tea is also good for treating diarrhoea, and nausea.

Figure 10: Ginger Tea

Other specialty ginger teas are ginger-lemon tea and ginger-honey tea. Lemon juice and ginger roots are used to prepare ginger-lemon tea and honey and ginger roots are used to prepare ginger-honey tea.

In South India, ginger powder along with black pepper powder is used to make a specialty coffee which is highly medicinal and used as a home remedy for treating cough, cold, sore throats, and common fevers. For preparing this medicinal coffee, firs of all, water is boiled and then coffee powder and sugar are added to make the base. To this coffee, a pinch each of ginger powder and pepper powder are added and the mixture is again boiled. This coffee should be consumed hot.

How to Prepare Ginger Ale: Ginger ale is a carbonated soft drink flavoured with ginger. Traditionally, ginger ale is made from a mixture of yeast, sugar, fresh ginger roots, other flavours as per the requirements and water. The carbonation process is the result of the yeast fermentation. Ethanol is a byproduct of fermentation and the ethanol percent in ginger ale can be controlled by modifying duration of the fermentation period.

Commercially, ginger ale is prepared with a mixture of carbonated water and sugar or high-fructose corn syrup. For flavouring purposes, either artificial or natural ginger-flavours are added.

How to Prepare Ginger Wine: Ginger wine is made from the fermentation process of the mixture containing ground ginger roots and raisins along with yeast and sugar. Ginger wine is often consumed with other cocktails such as lemonades and ginger ale.

How to Prepare Ginger Beer: Ginger beer is a brewed, naturally sweetened and carbonated non-alcoholic soft drink. Traditional ginger beer is prepared from the natural fermentation of ginger roots, yeast and sugar.

Health Risks: Fresh ginger paste may be applied as a topical ointment but it is not recommended because it may cause allergic reactions to some people. Ginger and ginger-based products should be used with great precaution during pregnancy.

Growing Practices for Ginger: It is easy to grow ginger plants. A detailed account of growing practices for ginger plants is given below:

Commercial Varieties: There are several ginger varieties available for cultivation. Major commercial varieties of ginger are 'China', 'Maran', 'Himachal', 'Rio-de-Janeiro', 'Nadia', 'Thingpuri', 'Narasapattam', 'Wynad Manantoddy', 'Karkal', 'Vengara', 'Ernad Manjeri', 'Ernad Chemad' and 'Burdwan'.

However in international spice trading markets, one can see ginger varieties named after the places of their production. In these markets, Jamaican ginger (ginger produced in Jamaica) is considered as the best variety. Then there is Kenya ginger which has also got good market. African and Indian ginger varieties which have comparatively darkish brown skin are considered inferior in quality. Another ginger variety is Japanese ginger.

Climate: Ginger is a shade-loving, tropical plant with preference for wet to moist areas; it cannot stand direct scorching sunlight, frost, very low temperatures, waterlogged soil and strong winds. Ideal climate for ginger cultivation is warm, humid, tropical and subtropical climate. It can be grown up to an altitude of 1500 meters from MSL (mean sea level).

Soil: Well-drained, rich and moist soil with plenty of organic matter is the most ideal soil for ginger cultivation.

Light: Ginger prefers filtered sunlight and hence grows well in partially shaded locations.

Crop Rotation: Ginger can be grown both as a rainfed crop and as an irrigated crop. In irrigated crop, ginger crop can be rotated with turmeric, low-growing vegetables such as onion and garlic, fruit trees such as plantain and banana, betel vine crop, chillies, groundnut, maize and sugarcane. In rainfed crop, ginger crop may be rotated with tapioca, sweet potato, yam and dry paddy once in 3 or 4 years.

Ginger as an Intercrop: Ginger crop may be raised as an intercrop among coffee, coconut, areca nut and orange plantations.

Ginger as a Mixed Crop: Ginger crop may be raised as a mix crop along with shade-giving plants such as plantain, banana, and tree castor.

Propagation: Propagation is by planting rhizome (modified underground stem) cuttings directly in the main field. Each rhizome cutting weighing 15-30 grams and having at least 2-3 well developed growth buds is planted.

Field Preparation: Field is prepared by ploughing twice or thrice followed by levelling and ridge making. It is always better to plant ginger rhizomes on the raised beds.

In tropical countries such as India, land preparation begins during March and April. Land is ploughed into a deep fine tilth and raised beds of 15 cm height and 1 meter width and of convenient length up to 5 or 6 meter are prepared. Soil solarization is done before planting as a preventive measure. Soil solarization process kills all plant pathogens present in the soil.

Planting for Rainfed Crop: A spacing of 30 cm is given between two ridges/raised beds. Shallow pits are prepared on the ridges to plant rhizome cuttings which are planted in two rows. Row to row distance is 30 cm and plant to plant distance within a row is 15-20 cm.

Planting for Irrigated Crop: A spacing of 45-50 cm is given between two ridges/raised beds. Shallow pits are prepared on top of the ridges in a single row to plant rhizome cuttings. Plant to plant distance within a row is 25 -30 cm.

Sowing Depth: Rhizome cuttings are normally planted approximately 5- 10 cm deep into the pits with the growing buds up and then it is covered by a thin layer of soil.

Figure 11: Rhizome Sprouts for Propagation

Rate of Planting Materials Required: Approximately 2 tons of rhizome cuttings are required to plant one hectare of land. That is, about 750-1000 Kg seed rhizomes are required to plant one acre.

Planting Time: Late winter or Early Spring (March to May) is the best time for planting ginger rhizomes.

Fertilizer and Manure Application: Ginger is a heavy feeder and hence requires heavy manuring. Application of 5-10 tons of farmyard manure or compost per acre at the time of field preparation is recommended. Subsequently, vermicompost or any other organic fertilizer available locally, may be applied @500 Kg/acre. Recommended N, P, K (nitrogen, phosphorous and potassium) fertilizer doses are 75:50:50 Kg/hectare. Whole phosphorous and half potassium fertilizers are applied at the time of planting. Remaining half potassium and half nitrogen fertilizers are applied two months after planting. Remaining half dose of nitrogen fertilizer is applied one month after second application of fertilizers. In some parts of the world, farm yard manure or neem cake or castor cake @4-5 t/ha is used in 2-3 top dressings.

Irrigation: For irrigated crop, first irrigation is done soon after planting; subsequent irrigations are given just to keep the soil moist throughout the growth phase of ginger plants. Water logged or soggy soils should be avoided. For rainfed crop, moist leaf mulch (of dried or green leaves) is spread over the beds soon after planting.   In certain areas, farm yard manure is used as mulch. New shoots emerge within 2-3 weeks after planting.

Mulching of the Ginger Beds: Mulching is an important cultural practice while growing ginger. At least two or three mulching are required during the growth period of a ginger plantation. First mulching is done soon after planting rhizome cuttings. Mulching with green leaves is highly recommended as this practice is proved to be more beneficial for ginger growth. Second mulching is done after first weeding and hoeing practices are over, which is about 40-50 days after planting. Third mulching is done after second weeding and hoeing practices are over, which is about 40-50 days after second mulching. Subsequent weeding and hoeing, and mulching practices are done as and when necessary.

Interculture and Aftercare: Hoeing and weeding are done to keep the field weed-free. It is recommended to mulch the beds thickly with biodegradable mulch soon after planting rhizome cuttings. This biomulch effectively controls weeds as well as conserves moisture.

Insect-Pest Management: Ginger is susceptible to spider mite attack in a dry weather; organic insecticides based on pyrethrum or tobacco extracts may be used to control them. Other major pests that are found attacking ginger are shoot borer (Conogethes punctiferalis/ Dicrhosis punctiferalis), nematodes, and white grub (Holotrichia setticolis). For controlling shoot borers, regular field surveillance is required which needs to be followed up by proper phyto-sanitary measures. Another option is hand picking of caterpillars and destroying them. Some growers grow neem trees in ginger plantations because of the insect-repellant properties of the neem trees.

For controlling nematodes, application of neem cake @1ton/ha is recommended. Two applications are required one during planting and second application 45 days after planting. White grub (Holotrichia setticolis) may be controlled by tillage of fields particularly during summer and by solarization of fields. Setting up of bird perches, other bird attractants and light traps and handpicking of infested leaves and grubs may also effectively control white grub infestation.

Disease Management: Soft rot or rhizome rot caused by *Pythium aphanidermatum* and *Pythium myriotylum* is a major disease found affecting ginger rhizomes. For effective control of this disease, the following cultural practices may be followed:

1. Selection of planting materials from disease free areas
2. Ensuring proper drainage of fields
3. Soil sterilization by solarization
4. Sanitation of fields by burning of infected plants
5. Removal of affected plants
6. Application of *Trichoderma viride* at the time of planting mixed with farm yard manure
7. Restricted use of popular fungicide Bordeaux mixture (1%) in the areas susceptible to this disease

Another major ginger disease is bacterial wilt caused by *Ralstonia solanacearum/ Pseudomonas solanacearum*. Selection of disease free planting materials and crop rotation of ginger with maize, cotton, and soybean may effectively control this disease.

Days to Maturity: It takes about 8-10 months for a ginger plantation to reach maturity.

Harvesting: Harvesting is done when the leaves have completely withered and the rhizomes are still tender and immature with outer skin still has a slight greenish colour. Rhizomes are carefully lifted by a digging-fork. Care is taken while lifting the rhizomes to avoid any sort of bruises and mechanical injuries of the rhizomes.

Figure 12: Freshly Harvested Ginger

Curing of Freshly Harvested Ginger Rhizomes: Curing is an important post-harvest practice in gingers. Freshly harvested ginger rhizomes are transported to pack houses where they are cleaned. All dirt and adhering fibrous roots are removed. Rhizomes with plump buds normally called as 'seed gingers' are segregated and stored for next planting season. Remaining commercially acceptable rhizomes are graded and packed before marketing them as 'fresh ginger'.

For dry ginger, fresh gingers are soaked in water to facilitate peeling process. Peeled rhizomes are dried in the sun before marketing them as 'dry ginger'.

Yield: Yield varies depending on the variety, soil fertility, cultural practices and prevailing climatic conditions. Approximately a ginger plantation may yield 10-15 tons of fresh ginger per hectare. Dry ginger recovery is about 20-30 percent depending upon the variety used.

Storage: Fresh ginger can be stored in a refrigerator or in cold storage for several weeks without losing its freshness. Dried and powdered ginger may be stored in airtight containers in a cool, dry place for a number of years.

Seed ginger is best stored by keeping them in pits which are dug in a cool place, away from sun and rain. Before storing them, they are treated with an effective fungicide as a preventive measure against soft rot and other soil-borne fungal infections. Seed gingers thus treated are then dried in shade and placed in pits which are about one meter in depth. A layer of sand or saw dust is placed in the pit before placing seed gingers and after placing the seed gingers, the pit is covered with a wooden plank while providing maximum aeration for the stored gingers.

In higher altitudes, seed ginger may safely be stored in an underground storage chamber until next planting season.

Storage Disorders: Rhizome rot (common among bruised rhizomes), shrivelling and drying of rhizomes, dry rot and sprouting of rhizomes are observed during storage.

Grading: Grading of ginger is mainly done depending on its dry matter content and fiber content. Ginger roots having highest dry matter and lowest fiber is regarded as the best grade or

Grade 1. Poorest graded ginger roots contain higher fiber content and low dry matter content.

Ginger as a Garden Plant: Ginger plant is very popular as a kitchen garden plant in many tropical and subtropical gardens. It is very easy to grow ginger plants in containers. Ginger plants are suitable for hydroponics or soil-less gardening also.

TURMERIC

Scientific name of turmeric is *Curcuma longa*. Synonym is *Curcuma domestica*. It belongs to the family Zingiberaceae, the ginger family. As in case of all plants belonging to ginger family, turmeric also prefers tropical and subtropical climate and a warm moist weather for its growth. Turmeric plant is an herbaceous perennial crop mainly grown for its edible rhizomes which are used as an important spice, condiment and dye. Turmeric is also known as 'Indian saffron'. Turmeric is a highly regarded medicinal spice in India and other South East Asian countries.

Figure 13: A Turmeric Plantation

Origin and Distribution: Turmeric is believed to be a native of South East Asia, particularly India and China. It is cultivated as a commercial crop in West Bengal of India, South India, China, Taiwan, Sri Lanka, Indonesia, Peru, Jamaica, Australia and the West Indies.

Figure 14: Turmeric Plant

Taxonomy: The taxonomic classification of turmeric is given below:

Kingdom	Plantae
Class	Equisetopsida
Subclass	Magnoliidae
Superorder	Lilianae
Order	Zingiberales
Family	Zingiberaceae
Genus	Curcuma
Species	longa

Botanical Description: Turmeric is a perennial herb growing from an underground rhizome with an erect unbranched stem. Stem is covered by leaf sheaths and grows up to a height of one meter upon full maturity. Leaves are long and lanceolate in shape just like ginger leaves but much larger and broader than ginger leaves. Leaves are bright green in colour. Yellowish flowers are borne in dense spikes. The most important plant part is rhizome, which is a modified underground stem with yellow flesh. Rhizomes are ready for harvesting within 8-10 months after planting.

Figure 15: Turmeric Rhizome

Economic Importance of Turmeric: Turmeric is an important spice and medicinal crop. Fresh turmeric roots are used for food purposes and as a traditional medicinal remedy for skin infections and various types of skin diseases. A large number of value-added products are prepared from turmeric. These are dried turmeric roots, turmeric powder, curcumin oil, turmeric Ayurvedic cream, and turmeric-based medicinal ointments.

Figure 16: Turmeric Powder

Food Uses of Turmeric: There are mainly two forms of whole turmeric available in the market: fresh turmeric and dried turmeric. Dried turmeric is used to prepare turmeric powder. Turmeric powder has a warm flavour and earthy, pleasant aroma. Turmeric powder is a major ingredient in many curry powder preparations. In almost all Indian food preparations, turmeric powder is used as a culinary dye to impart its specific yellow colour to food. It is believed that turmeric powder is rich in antioxidant curcumin which can prevent cancerous tumour formations in the cells.

Figure 17: Dried Turmeric

Curcumin oil, the essential oil extracted from fresh and dried rhizomes of turmeric is rich in curcumin and is used in perfume industry and also in food manufacturing industry.

According to USDA, turmeric spice offers a lot of nutritional and medicinal benefits. Actually, turmeric powder is about 67-70% carbohydrates, 10-13% water, 9-10% protein, 3-4% fat/lipid, 3–7% dietary minerals, 3–7% essential oils, 2–7% dietary fiber, and 1–6% curcuminoids. Major curcuminoids are curcumin, demethoxycurcumin, and bisdemethoxycurcumin. Curcumin constitutes up to 3% of commercial-grade turmeric powder. Curry powder made from turmeric powder contains an average of 0.3% curcumin. Major essential oils present in turmeric are turmerone, germacrone, atlantone, and zingiberene. A detailed account of the nutritive value of turmeric powder is given below:

Nutrient	Unit	Value/100 g
Water	g	12.85
Energy	kcal	312
Protein	g	9.68
Total lipid (fat)	g	3.25
Carbohydrate	g	67.14
Fiber, total dietary	g	2.7
Sugars, total	g	3.21
Calcium, Ca	mg	168
Iron, Fe	mg	55
Magnesium, Mg	mg	208
Phosphorus, P	mg	299
Potassium, K	mg	2080
Sodium, Na	mg	27
Zinc, Zn	mg	4.5
Vitamin C	mg	0.7
Thiamin	mg	0.058
Riboflavin	mg	0.15
Niacin	mg	1.35
Vitamin B-6	mg	0.107
Folate, DFE	µg	20
Vitamin B-12	µg	0
Vitamin A, IU	IU	0
Vitamin E (alpha-tocopherol)	mg	4.43
Vitamin K (phylloquinone)	µg	13.4
Fatty acids, total saturated	g	1.838
Fatty acids, total monounsaturated	g	0.449
Fatty acids, total polyunsaturated	g	0.756

Medicinal Uses of Turmeric: Turmeric is anti-cancerogenic (fights against cancer-causing cells), antiseptic (kills harmful microbes), antifungal (cures fungal infections) and antiviral (cures viral infections). The active ingredient that imparts turmeric its medicinal properties is called *curcumin* which is present in both fresh and dried rhizomes of turmeric. Turmeric is also consumed internally as a stimulant.

Turmeric is anti-inflammatory and curcumin is a powerful antioxidant. Food rich in antioxidants are good as an anti-ageing diet. Antioxidants are also helpful in fighting cancerous cells. Anti-inflammatory property of curcumin is used in treating arthritis.

Turmeric is antifungal and antiseptic because of the presence of curcumin. Hence turmeric paste may be applied to cure leech bites, ring worms, skin inflammations, mouth sores, wounds, insect bites etc.

Growing Practices for Turmeric: It is very easy to grow turmeric plants. Growing practices of turmeric plants are almost similar to those of ginger plants. A detailed account of various growing practices for turmeric is as given below:

Commercial Varieties: There are a number of turmeric varieties available for cultivation. Some of the major varieties of turmeric are 'Aleppey Turmeric', 'Madras Turmeric', 'Lokhandi', 'Duggirala', 'Tekurpeta', 'Kasturi Pasupa',

'Armoor', 'Roma', 'Suroma' and 'Chaya Pasupa'. Turmeric varieties with more than 5% curcumin content and having lemon yellow or orange yellow interior are preferred in the international markets.

There is a highly aromatic variety of turmeric available in the market which is called 'kasturi turmeric'. Its scientific name is *Curcuma aromatica*. It has a great cosmetic value and used extensively in cosmetic industries for preparing face creams, facial scrubs and face wash.

Climate: Turmeric needs a warm and humid climate. It can be grown as a rainfed crop in heavy rainfall areas. In other areas it is grown as an irrigated crop. It can be successfully be grown from mean sea level (MSL) up to an altitude of 1200 meters.

Soil: Turmeric crop thrives well in well-drained fertile loamy soils which are rich in humus content. Turmeric plantation cannot withstand waterlogged soils and soil alkalinity.

Propagation: Propagation is mainly via rhizome cuttings and rhizome fingers (small tubers attached to mother rhizomes).

Crop Rotation: Crop rotation is recommended because turmeric is an exhaustive crop just like ginger. In wet and moist lands, turmeric crop may be rotated with paddy, sugarcane, banana and plantain once in 3-4 years. In plains turmeric can be grown in rotation with sugarcane, chilli, onion, garlic, wheat, pulses and short season vegetables.

Turmeric as an Intercrop: In tropics, turmeric may be grown as an intercrop with mango, coconut, areca nut, jackfruit tree and litchi plantations.

Field Preparation: Land is ploughed 3-4 times to bring the soil to a fine tilth. Then raised beds or ridges of one meter width and length of convenient size with a height of 15 cm are prepared with a spacing of 30-45 cm between two ridges.

Planting for Rainfed Crop: A spacing of 30-45 cm is given between two ridges/raised beds. Shallow pits are prepared on the ridges to plant rhizome cuttings which are planted in two rows. Row to row distance is 30 cm and plant to plant distance within a row is 15 cm.

Planting for Irrigated Crop: A spacing of 45-60 cm is given between two ridges/raised beds. Wider spacing is allowed to facilitate irrigation process. Shallow pits are prepared on top of the ridges in a single row to plant rhizome cuttings. Plant to plant distance within a row is 15-25 cm.

Planting Materials: Both the mother rhizomes and fingers are used as planting materials. Mother rhizomes are planted as such or split into two or more parts and used for planting. Fingers are cut into 5 cm long cuttings along with at least one healthy bud in order to be used as planting materials.

Sowing Depth: Rhizome cuttings or fingers are normally planted approximately 5- 10 cm deep into the pits prepared on the ridges with the growing buds up and then it is covered by a thin layer of soil.

Rate of Planting Materials Required for Planting: About 2-3 tons of rhizomes and fingers are required to plant one hectare of land.

Planting Time: In tropics, turmeric is planted during April to July.

Fertilizer and Manure Application: Turmeric plants require heavy manuring. Chemical fertilizers are seldom used in turmeric plantations. Turmeric is mostly grown with organic manures and biofertilizers. First application of organic manures i.e. FYM (farm yard manure) or compost is done at the time of field preparation@4-5 tons/hectare. Then soon after planting, groundnut cake or neem cake is applied @1-2 tons/hectare in two equal doses. First application is 2 months after planting and second application is done 2 months after first application. Vermicompost or coir pith compost or any other organic manure may be supplemented with the application of compost or FYM.

Irrigation: Irrigation is done in the furrows available between two ridges/raised beds. For irrigated crop, first irrigation is done soon after planting; subsequent irrigations are given just to keep the soil moist throughout the growth phase of the plants. Water logged or soggy soils should be avoided. For rainfed crop, moist leaf mulch (of dried or green leaves) is spread over the beds soon after planting. In certain areas, farm yard manure is used as mulch. New shoots emerge within 2-3 weeks after planting.

Mulching of the Turmeric Beds: Mulching is an important cultural practice while growing turmeric. At least two or three mulching are required during the growth period of a turmeric plantation. First mulching is done soon after planting rhizome cuttings. Mulching with green leaves is highly recommended as this practice is proved to be more beneficial for turmeric growth. Second mulching is done after first weeding and hoeing practices are over, which is about 40-50 days after planting. Third mulching is done after second weeding and hoeing practices are over, which is about 40-50 days after second mulching. Subsequent weeding and hoeing, and mulching practices are done as and when necessary.

Interculture and Aftercare: Hoeing and weeding are done to keep the field weed-free. It is recommended to mulch the beds thickly with biodegradable mulch soon after planting rhizome cuttings. This biomulch effectively controls weeds as well as conserves moisture.

Insect-Pest Management: Shoot borer is a major insect found attacking turmeric plantations. In order to control shoot borers the infected shoots may be removed and destroyed. Neem oil

5% spray once every 2 weeks is found to be effective in controlling shoot borers.

Disease Management: Leaf spot and leaf blotch are two major diseases of turmeric. These fungal infections may be treated successfully by the restricted application of fungicide Bordeaux mixture (1%). Another major disease is rhizome rot. It can be controlled either by soil solarization or by *Trichoderma* application to soil at time of planting.

Harvesting: Turmeric crop will be ready for harvesting within 8-10 months after planting. Turmeric is harvested when green leaves have completely turned yellow and started withering. Dried leaves are cut close to the ground. Then land is irrigated if necessary to facilitate harvesting process. The rhizomes are dug and taken out by using a digging fork or crowbar or a spade. Alternatively, the rhizomes may be hand-picked after ploughing the land.

Figure 18: Freshly Harvested Turmeric Rhizomes

Curing Turmeric: Freshly harvested turmeric rhizomes are transported into pack houses where they are cleaned from dirt and all attached fibrous roots are removed. Fingers are separated from mother rhizomes and kept for curing before selling in the markets. Mother rhizomes are normally used as planting materials for next season planting.

Finger turmeric is boiled in drums until white fumes appear giving out characteristic earthy turmeric odour. Then the cooked fingers are spread on a clean floor or mat in the sun in order to dry them. Generally two weeks of sun drying is required to obtain properly cured turmeric fingers. The market of turmeric depends on its curcumin content and colour. Proper curing process is essential in order to ensure proper colour development and quality in the cured turmeric.

Storage: Fresh turmeric may be stored in a refrigerator or in cold storage for several weeks. Turmeric powder if stored in an air tight container in a cool place will last for several years. Cured dried turmeric can be stored in a dry cool place in gunny bags.

Yield: Approximately 20-25 tons of turmeric is obtained from one hectare turmeric plantation. Cured turmeric is about 25 percent of the fresh rhizomes by weight.

Figure 19: Turmeric Flower

INDIAN ARROWROOT

Scientific name of Indian arrowroot is *Curcuma caulina*. Synonym is *Hitchenia caulina*. It is a tall herbaceous perennial plant belonging to ginger family i.e. Zingiberaceae. It is also known as arrowroot lily. In Hindi language it is called *tikhur*. The plant grows up to a height of 50 cm to 1 meter upon full growth. The plant has perennial rhizomes with numerous hanging tubers. The rhizomes are used for extracting edible starch.

Origin and Distribution: Indian arrowroot is believed to be a native of South-Western deciduous forests of India, more precisely speaking the Western Ghats of Maharashtra and surrounding regions. In Marathi language, it is called *chavar*. Indian arrowroot is found wildly growing along the moist forests of Western Ghats. In some parts of South India, Indian arrowroot is commercially cultivated as a root crop just like ginger and turmeric for extracting arrowroot powder. Arrowroot powder is believed to have numerous medicinal properties. Arrowroot powder is also used to make edible starch and arrowroot biscuits.

Taxonomy: Taxonomic classification of Indian arrowroot is as given below:

Kingdom	Plantae
Class	Equisetopsida
Subclass	Magnoliidae
Superorder	Lilianae
Order	Zingiberales
Family	Zingiberaceae
Genus	Curcuma
Species	caulina

Economic Importance: Indian arrowroot is mainly grown for its rhizomatous tubers which have both food and medicinal uses. Tubers, which are actually modified rhizomes, are rich source of starch. Tubers are just like turmeric and ginger tubers in appearance but with white flesh. Commercially important arrowroot powder is prepared from dried tubers of Indian arrowroot. Arrowroot powder is extensively used in Indian traditional medicines. The plant may be planted in high rainfall areas along the sides of rivers and irrigation canals to prevent soil erosion.

Actually the word, "Arrowroot" is really confusing. Indian arrow root should not be confused with *Maranta arundinacea*, which is also known as 'arrowroot plant' because of its starchy tubers. In fact, *Maranta arundinacea* is known as "True Arrowroot" or "West Indian Arrowroot". Sometimes it is simply known as "Maranta" and tubers of which are also used in the preparation of arrowroot powder. Another close relative of "Indian Arrowroot" is "East Indian Arrowroot" plant or

Curcuma aungustifolia which is recognized as a medicinal herb.

Extraction of Arrowroot Powder: Freshly harvested tubers are transported to the pack houses where they are washed before cleaning them by removing all adhering fibrous roots and small tubers. Cleaned tubers are then grated to obtain white coloured pulp which is again washed thoroughly before sieving process. Sieved pulp is again washed and kept for hours for the starch to settle out. The starch is extracted and dried and packed for further use.

Botanical Description: Indian arrowroot plant is an herbaceous perennial plant growing from an underground rhizome with a height of 50 cm to 1m. The plant is erect-growing with unbranched leafy stems. Leaves are oblong lanceolate in shape which reach up to 50 cm in length on full growth. Yellowish or pinkish white or greenish white flowers are borne on a spike which is 10-15 cm in length. Flowers are attractive, and highly ornamental in appearance. Flowering time is July to October in tropics.

Growing Practices for Indian Arrowroot: It is very easy to grow these plants. A detailed account of growing practices for Indian arrowroot is given below:

Climate: Indian arrowroot is a plant of hot tropics with a preference for moist to wet areas. Just like ginger and turmeric, Indian arrowroot also likes moist, shady places for its growth.

It cannot stand direct scorching sunlight, frost, very low temperatures, waterlogged soil and strong winds. The plant grows well in high rainfall areas with the mean annual rainfall is more than 5,000mm.

Soil: Well-drained, rich and moist soil with plenty of organic matter is the most ideal soil for the cultivation of Indian arrowroot.

Light: Indian arrowroot prefers filtered sunlight and hence grows well in partially shaded locations.

Propagation: Propagation is mainly through cuttings of the rhizomes.

Planting Time: Indian arrowroot is planted at the onset of monsoon.

Harvesting Time: For commercial cultivation, Indian arrowroot plants are cultivated as biennials where first year vegetative growth is promoted and in the second year tubers are harvested when they are 20-24 months old.

Harvesting: Harvesting is done when the leaves and stalks have completely withered.

Yield: Yield varies depending on soil fertility, cultural practices and prevailing climatic conditions.

CONTAINER GARDENING PRACTICES

The practice of growing plants in pots and containers is known as container gardening. Container gardening is mostly practiced in urban homes, multistorey buildings of towns and cities and in places where availability of land is a major constraint. All members of ginger family such as ginger, turmeric and Indian arrowroot are popular container-grown plants among urban populations. They can be grown in containers successfully.

Figure 20: Container Gardening of Turmeric

Some of the major considerations while preparing for raising a container garden are: choosing right containers, choosing a good growing media, and preparing a gardening schedule or a garden calendar.

Selection of Suitable Containers/Pots: Earthenware pots are most commonly used containers for growing gingers. Wooden barrels and planters can also be used and these containers should be painted from inside as well as outside with waterproof oil-paints before using them. Plastic jars, pots, dishes and bowls, glazed clay and china (porcelain) pots, shallow bowls and troughs, pottery containers, boxes and crates, cement pots, cans and buckets, tin boxes, drums, brass, copper containers, or any such suitable containers may be used according to the circumstances and growing requirements of the gardeners. Similarly, when it comes to the shape of the containers, any shape of the container can be used whether it is circular, round, oval, elliptical, cone, pyramid, rectangular, square, or heart-shaped.

Figure 21: Terracotta Pots for Container Gardening

Containers should have at least one hole of an adequate size at the bottom as in earthen pots, to drain out excess water. Containers should easily be placed on the terrace, window sills, window boxes, balcony and verandah where sunlight is available for the plants. Containers should be able to hold sufficient volume of growing media and should be lightweight (for portability, if needed), and easy to handle. Containers should be durable and free of toxic substances and also should prevent root circling.

Selection of Suitable Growing Media or Soil and Fertilizers: Growing media may be prepared from a mixture of good soil, river sand, well-decomposed organic manure (compost or farmyard manure), nitrogenous fertilizers (urea or ammonium sulphate) and recommended organic insecticides and fungicides.

Growing media should be able to hold seedlings firmly. Media should be free of insect-pests, and should have good water-holding capacity. Growing media should also have excellent aeration and drainage. If it is inconvenient to prepare growing media at home, one can purchase ready-made growing media from a plant nursery.

How to prepare a suitable growing media? Mix good soil, river-sand and well-rotten organic manure in equal quantities with the help of a shovel. Make sure that the mixture is free from various soil-borne insects, termites, red ants and cut worms. Add a small quantity of recommended fungicide (organic pyrethrum-based fungicides may be used)to the mixture before filling it in the containers; this helps to prevent seedling rotting caused due to fungal infections. Usage of synthetic chemicals is not advised for container grown or indoor grown plants due to chemical-residue related hazards and health risks.

Figure 22: Preparation of Growing Media

After raising a crop for one season, the container mixture should be removed and cleaned of roots and exposed to the sun for a few days. This growing media could then be reused after mixing one-third the quantity of organic manure and a small quantity of recommended fungicide. Alternatively, growers can prepare their own compost or vermicompost by using kitchen wastes and use it as growing media for their container-grown plants.

Preparation of a Garden Calendar: If you are planning for container-growing of gingers, then you should prepare a gardening schedule beforehand. A garden calendar should clearly address the following issues:

1.   Sowing Time: When to sow the seeds?

2. Fertilizer Schedule: When to fertilize plants? How many times?

3. Irrigation/Watering Schedule: How much irrigation is required? And when?

4. Weeding and Aftercare: What are the weed control measures to be adopted?

5. Harvesting Time: When can the leaves be harvested?

In addition to this, a container gardener should arrange the following garden tools:

- Shovel: A shovel is needed for mixing the soil-manure mixture.
- Hand Cultivator: It is needed for working around plants and breaking up soil clods for light replanting.
- Trowel: It is used for transplanting purposes.
- Watering Can: A watering can is required for watering plants.

Figure 23: Watering Can

- Hand Duster: It is used to apply chemicals in powder form.

- Garden Gloves: These are needed to protect the gardener's hands.

Figure 24: Garden Gloves

- Air Sprayer: It is the most popular equipment for

applying liquid fertilizers, pesticides and fungicides through foliar spraying because it gives good coverage.

Figure 25: Sprayer

- Strings and Measuring Tapes: These are required for measuring purposes.
- Wheelbarrow or Garden Cart: It is used to transport soil and manures, garden tools, and harvested produce
- Seeders: It is used for sowing seeds.

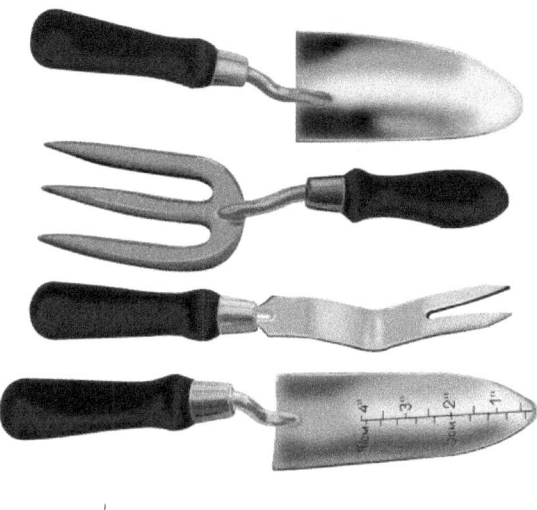

Figure 26: Garden Tools for Container Gardening

How to Grow Gingers in Containers? After arranging containers, growing media and necessary garden tools, a grower can start raising seedlings.

Rhizome cuttings may be planted directly into the containers. Alternatively, planting materials may be raised in well-prepared nursery beds. Nursery beds should be prepared in a well-shaded area. Soil should be well-drained and pathogen-free. Regular light watering is necessary to keep the nursery bed moist always. Moist soil facilitates quick root development in rhizome cuttings. Well-developed seedlings of 10-15 cm height are later transplanted in the containers.

Remember that plants in pots and containers need a lot of care and attention. So it is essential to water and manure frequently the growing plants depending on the climate, size of the plant and type of container. Plants need extra water in dry summer season because of evaporation/transpiration losses, so watering should be done twice a day (morning and evening). Too much watering can be as harmful in winter as too little in summer. In the rainy season, proper water drainage is essential if plants are placed in the open areas. If there is heavy rain, containers should be tilted slightly to drain out the excess water from the top.

Topdressing with nitrogenous fertilizers improves plant growth and fruit yield. This can be done by foliar application of liquid nitrogen, urea or ammonium sulphate in small quantities. Alternatively, urea granules@ 5–10g/container may be applied in moist soil once a week or 10 days, starting from 2 weeks after transplanting the seedlings. *High dose of fertilizer is very harmful since it can kill the plants.*

If urea or ammonium sulphate is applied in dry soil, the plants must be watered immediately. Young plants may require staking. Hand-hoeing and weeding with the help of a small shovel should be done periodically to remove weeds, if there are any. Weeds may also be uprooted gently by hand.

For container-grown plants, water requirement may be determined by weighing pots, feeling growing medium and by using indicator plants that readily show water stress. Watering should be done in early morning to minimize evaporation loss. Applying water in two or more applications conserves water.

Major insects found in container-grown plants are, aphids, mites, mealy bugs, ants, and leaf borers. Aphids and jassids are small-sucking insects, injuring the plants especially in early stage of their growth. Leaf borers damage young leaves and make them unfit for consumption. Use of organic insecticides such pyrethrum-based sprays or tobacco emulsions, or neem oil based solutions effectively control these insects. Sometimes, spraying with diluted soap solution will also give good results. After spraying with insecticides, rhizomes/roots should not be harvested for 7 days for consumption as insecticidal residues may be present in them. Use of mechanical traps (colour traps, light traps etc) and manual picking of insects may also be tried for insect control.

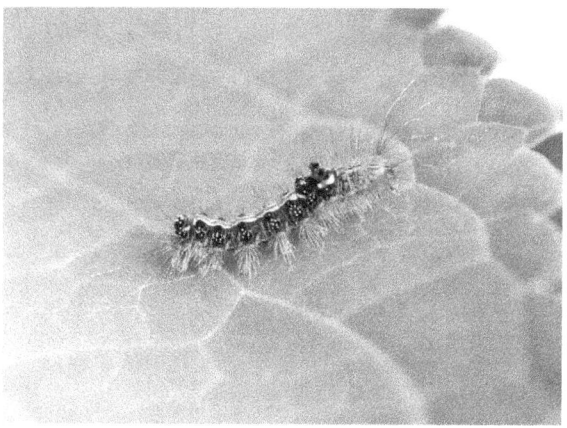

Figure 27: Leaf-Eating Caterpillars

Fungal diseases (damping off and wilt) and viral diseases affect the plants kept in the open areas, particularly in the rainy season. Fungal diseases can be controlled by drenching the soil with an appropriate fungicide. Virus-affected plants should be removed and destroyed.

Root pruning is essential to remove root circling when root systems become too long for their containers. Root circling can be prevented by air root pruning or by using bottomless containers and copper compounds. Pruning may be necessary to induce new root growth.

Repotting: Repotting the plants into bigger containers may also be tried to accommodate overgrown plants.

Growing Ginger and Turmeric in Polybags: Ginger, turmeric, and other members of ginger family are suitable for growing in polybags and jute bags. However only one crop can be raised by using such bags.

ORGANIC GROWING PRACTICES

Growing plants organically is becoming a healthy trend nowadays because of the eco-friendly growing practices adopted in organic agriculture. The products of organic agriculture are extremely safe to consume without fearing any dangers of pesticidal residues.

The key areas of organic agriculture is the use of seeds/planting materials of organic origin, the use of organic manures, compost, green manures, and biofertilizers for plant nutrition, adopting integrated pest management technology for controlling insects and pests and integrated disease management for controlling plant diseases and also using a lot of beneficial cultural practices such as crop rotation, companion planting, using trap crops, mulching, soil solarization, soil fumigation, etc for maintaining a healthy ecosystem in the growing fields. Organic farming technology focuses on using biocontrol agents and natural enemies/predators and biopesticides for effective control of insects and pests.

Thus, in a nut shell, major steps involved in organic growing of ginger and turmeric are:

1. Choosing planting materials of organic origin
2. Preparing soil and enhancing soil fertility by adding compost, biofertilizers and organic manures
3. Integrated Pest Management
4. Integrated Weed Management
5. Integrated Disease Management
6. Other Cultural Practices such as Soil Solarisation, Crop Rotation, Mulching, Companion Planting/Intercropping, Using Trap Crops etc

Choosing Plant Materials: In organic production, we need to use seeds and planting materials of organic origin.

Preparing Soil: For growing any plants, as we all know, soil is the major growing medium. So soil should be fertile with all essential plant nutrients to support the healthy plant growth. Generally, soil fertility may be enhanced by the addition of organic manures and organic fertilizers into the top soil.

In organic farming, seedlings/planting materials are planted in a sterilized soil medium so that infestation of soil-borne plant pathogens can be prevented. Soil sterilization can be done by soil solarization method. Soil solarization is a method of trapping solar energy within the soil by covering the soil with a transparent polyethylene cover for a certain period of time.

This practice kills all soil-borne plant pathogens.

After soil sterilization, methods need to be adopted to enhance soil fertility. This is accomplished through the addition of organic manures, compost, vermicompost, biofertilizers, and organic fertilizers.

Now, what is a fertilizer? Fertilizer is the term used to refer any material that provides essential nutrients for plant growth when applied externally, either by mixing with soil or by dissolving in water. A fertilizer is rich in essential plant nutrients. Fertilizer may be either in solid form or in liquid form.

Fertilizer application is an important cultural practice while growing plants. It is the practice of adding fertilizers to the soil or any other growing media through direct broadcasting; or by direct mixing with the soil or through spraying of water-soluble fertilizers.

There are three types of fertilizers and these are Organic Fertilizers, Biofertilizers and Inorganic Fertilizers or Chemical/Synthetic Fertilizers. *In organic farming practices, we do not use chemical fertilizers.*

While growing plants organically, we generally use organic manures and organic fertilizers as well as biofertilizers. Any organic matter that may be used as an organic fertilizer is known as organic manure. Manures contribute to the fertility of the soil by adding organic matter and nutrients. There are three main classes of organic manures used in soil management: animal manures; compost and plant manures or green manures.

Animal Manures: Common forms of animal manure include farmyard manure (FYM) or farm slurry (liquid manure). FYM also contains plant material (often straw), which has been used as bedding for animals and has absorbed the feces and urine.

Figure 28: Farm Yard Manure/Cow Dung

Agricultural manure in liquid form, known as slurry, is produced by more intensive livestock rearing systems where concrete or slats are used, instead of straw bedding.

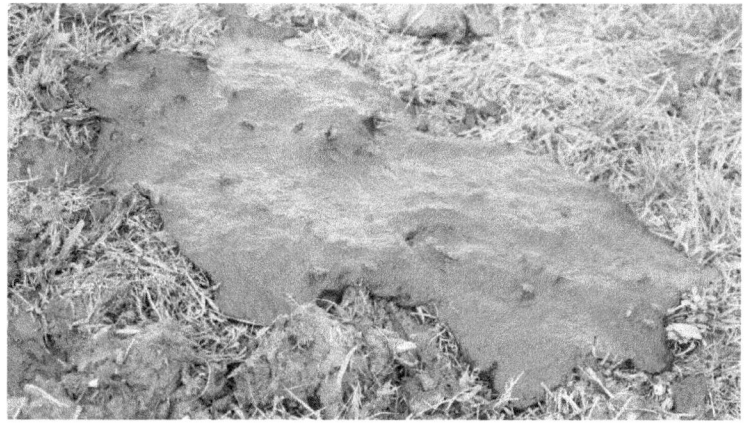

Figure 29: Farm Slurry

Sheep manure is high in nitrogen and potash while animal manures like pig manures are relatively low in nitrogen and potash. Horse manure may contain lots of weed seeds, as horses do not digest seeds the way that cattle do. So using horse manure may result in a lot of weed infestation in the fields. Chicken manure, even when well rotted, is very concentrated and should be used sparingly for growing plants. Animal manures may also include other animal products, such as wool shoddy (and other hair), feathers, blood and bone etc.

Compost: Compost is a decomposed plant matter that is used as an organic fertilizer, soil amendment and as a source rich in humus. Compost is used in organic farming, gardening, landscaping, horticulture, and agriculture. It may be used as a natural pesticide for soil and is useful for soil erosion control, soil reclamation, and waste land management. Composting is a natural microbiological process of decomposing organic matter into humus and minerals. Microorganisms that aid composting process include bacteria, actinomycetes, fungi, protozoa and earthworms. Composting process can be accelerated by shredding the leaves and adding extra nitrogen on the shredded leaves. Adding water also hastens composting process. Proper aeration must be ensured by regularly turning the mixture. The composting process is entirely dependent on micro-organisms to break down organic matter into compost. Presence of a healthy microbial community is essential for rapid decomposition process. Lack of a healthy microbial community makes composting process very slow. With the proper mixture of water, oxygen, carbon, and nitrogen, micro-organisms work faster to break down organic matter to produce compost. Shredding leaves make a homogenous compost mixture with smaller particles. Smaller particles decompose faster as there are more surfaces for the microbes to work on. Particles should not be very small as very small particles may compact and restrict oxygen availability. A blend of small and large particles will be most efficient. Compost can be prepared by using traditional process or by using modern composting techniques. Traditional composting is a slow process and it takes about one year or more to finish the composting process. It is simple, sometimes requires simply piling up waste outdoors or in pits. Modern composting uses containers or composting bins for composting process. It is a multi-step, closely monitored process and it uses more homogenized pieces in the compost. It uses measured inputs of water, air and carbon- and nitrogen-rich materials and is a rapid process. Modern composting process takes about 2 to 3 weeks to complete.

Figure 30: Compost

Major considerations while preparing compost are: Carbon, Nitrogen and Oxygen. Microbial oxidation of carbon produces the heat required for decomposition process. Nitrogen is necessary as microbes grow and reproduce consuming nitrogen-rich plant materials. Oxygen is also an important element as microbes oxidize carbon using the oxygen present in the compost pile. Water should also be added as presence of water maintains life activities of microbes within the compost pile. Optimum temperature should be maintained throughout the composting procedure. Higher temperature may kill the microbes. At low temperatures, they may remain inactive. The most efficient composting occurs with a carbon: nitrogen mix of about 30 to 1. Plant and animal materials have both carbon and nitrogen. If nitrogen is less in a compost pile, use urea fertilizer or other nitrogen-rich materials.

Figure 31: Composting Process

Standard uses are: 1 lb. urea to 1 cubic yard. Leaves, 6 lb. urea to 1 cubic yard. Wood clippings, 5 parts leaves to 1 part manure and dried blood meal, alfalfa meal at the rate of 2 cups to a wheelbarrow load of brown leaves or other carbon rich wastes such as shredded paper.

Moisture level should always be maintained at about 50%. Compost pile should always be moist and should never be kept dry. Overwatering should be avoided. Microbes need sufficient oxygen for decomposing compost. Since composting is an aerobic process, absence of oxygen causes anaerobic conditions causing a bad odor and partial decomposition of the compost. Hence it is essential that compost pile must be turned at regular intervals to facilitate aeration. Restrict size of the pile to no more than 5 ft. high and 5 ft. wide. Compost pile of 4x4 ft makes an ideal size. Optimum temperatures are between 100o and 140oF as higher temperatures than that may kill microbes. Composting at the center of the pile is complete when temperatures within the pile drop below 100o F. Once composting at the center is complete, turn the pile, putting outside edges inside and allow it to compost more.

Good compost will be dark in color; friable and porous and it will have an earthy smell. It is an excellent source of humus and plant nutrients and has good water holding capacity. Compost is used as an additive to soil and is used as a tilth improver, supplying humus and nutrients. It provides a rich growing medium and acts as a porous, absorbent material that holds moisture and soluble minerals, providing the support and nutrients in which plants can flourish. Compost may be mixed with soil, sand, grit, bark chips, vermiculite, perlite, or clay granules to produce loam.

Vermicompost: Vermicompost is the product of composting utilizing various species of worms. Red wigglers, white worms, and earthworms are used for vermicomposting. Vermicompost is a heterogeneous mixture of decomposed vegetable or food waste, bedding materials, and Vermicast. Vermicast, also known as worm castings, worm humus or worm manure, is the end-product of the breakdown of organic matter by species of earthworms. Red wigglers are recommended by vermiculture experts as they voraciously feed on compost pile and breed very quickly. Vermicompost contains water-soluble nutrients and is a nutrient-rich organic fertilizer and soil conditioner.

Figure 32: Vermicomposting

Green Manures: Green manures are crops grown for the express purpose of ploughing them in, thus increasing fertility through the incorporation of nutrients and organic matter into the soil. Leguminous plants such as clover are often used for this, as they fix nitrogen using Rhizobial bacteria in specialized nodes in the root structure. Leguminous cover crops are also grown to enrich soil as a green manure through nitrogen fixation from the atmosphere as well as phosphorus (through nutrient mobilization) content of soils. Other types of plant matter used as manure include the contents of the rumens of slaughtered ruminants, spent hops (left over from brewing beer) and seaweeds.

Figure 33: Clover, Green Manure Crop

What Are Organic Fertilizers? Organic fertilizers are naturally occurring fertilizers and mineral deposits. Organic manures are just one form of organic fertilizers. Other examples of organic fertilizers are vermicompost or worm castings, compost, seaweed, guano, naturally occurring mineral deposits (e.g. saltpeter) etc. Processed organic fertilizers include compost, blood meal, bone meal, humic acid, amino acids, and seaweed extracts. Other examples are natural enzyme digested proteins, fish meal, and feather meal. Decomposing crop residue (green manure) is also used as an organic fertilizer. Mined powdered limestone, rock phosphate and sodium nitrate, are inorganic (not of biologic origins) compounds but are approved for usage in organic agriculture in minimal amounts.

Organic fertilizer nutrient content, solubility, and nutrient release rates are typically much lower than mineral (inorganic) fertilizers. All organic fertilizers are classified as 'slow-release' fertilizers, and therefore cannot cause nitrogen burn. Organic fertilizers are low-cost as compared to inorganic fertilizers. An organic fertilizer improves the biodiversity (soil life) and long-term productivity of soil and also increases the abundance of soil organisms by providing organic matter and micronutrients for organisms such as fungal mycorrhiza, which aid plants in absorbing nutrients.

Initially organic fertilizers may not be as effective as inorganic fertilizers but application of organic fertilizers become as effective as chemical fertilizers over longer periods of continuous use.

Some of the major advantages of organic fertilizers are as follows:

- Nitrogen supplying organic fertilizers contain insoluble nitrogen and act as a slow-release fertilizer
- Increase physical and biological nutrient storage mechanisms in soils, mitigating risks of over-fertilization

- Mobilize existing soil nutrients, so that good growth is achieved with lower nutrient densities while wasting less

- Release nutrients at a slower, more consistent rate, helping to avoid a boom-and-bust pattern

- Help to retain soil moisture, reducing the stress due to temporary moisture stress

- Improve the soil structure

- Help to prevent topsoil erosion

Some of the major disadvantages of organic fertilizers are as follows:

- Organic fertilizers may contain pathogens and other disease causing organisms if not properly composted

- Nutrient contents are very variable and their release to available forms that the plant can use may not occur at the right plant growth stage

- Organic fertilizers are comparatively voluminous and can be too bulky to deploy the right amount of nutrients that will be beneficial to the plants

- The nutrients in organic fertilizer are both more dilute and also much less readily available to plants

- As a dilute source of nutrients when compared to inorganic fertilizers, transporting large amount of fertilizer incurs higher costs, especially with slurry and manure

- The composition of organic fertilizers tends to be more complex and variable than a standardized inorganic product

- More labor is needed to compost organic fertilizer, thus increasing labor costs

- Reduce external inputs of pesticides, energy and fertilizer, at the cost of decreased yield

Plant Nutrition: Plants absorb nutrients from the soil or the atmosphere, or from water. Carbon and oxygen are absorbed from the air while other nutrients including water are obtained from the soil. Three ways of nutrient uptake are Simple Diffusion, Facilitated Diffusion and Active Transport. Plants absorb essential elements from the soil through their roots and from the air (mainly consisting of carbon and oxygen) through their leaves. There are 17 essential plant nutrients grouped into two categories: macro-nutrients and micro-nutrients.

Major macro-nutrients are carbon, hydrogen, oxygen, nitrogen, phosphorus, potassium, calcium, magnesium, sulphur and silicon. Major micro-nutrients are boron, copper, chlorine, iron, manganese, molybdenum and zinc. Trace elements like sodium, nickel and cobalt may also be needed in certain circumstances.

Macronutrients are taken by plants in large quantities and are present in plant tissue in quantities from 0.2% to 4.0% dry weight while micronutrients are needed in small quantities and are present in plant tissue in quantities measured in parts per million, ranging from 5 to 200 ppm, or less than 0.02% dry weight. In the absence of an essential element, the plant is unable to complete a normal life cycle. An essential element is a part of some essential plant constituent or metabolite. An element present at a low level may cause deficiency symptoms, while the same element at a higher level may cause toxicity. Deficiency of one element may present as symptoms of toxicity from another element. An abundance of one nutrient may cause a deficiency of another nutrient. A lowered availability of a given nutrient may affect the uptake of another nutrient. The root, especially the root hair, is the most essential organ for the uptake of nutrients.

Now let us have a look at various functions of plant nutrients in detail...

- Carbon: It is backbone of many plants biomolecules, including starches and cellulose and is a part of the carbohydrates that store energy in the plant
- Hydrogen: It is necessary for building sugars and building the plant and is obtained from water. It is necessary for electron transport chain in photosynthesis and for respiration
- Oxygen: It is necessary for cellular respiration
- Nitrogen: It determines green color and density in plant and is needed for chlorophyll, which is needed for photosynthesis. It also improves plant's ability to

resist disease and tolerate effects of heat, cold, and drought. Major deficiency symptom includes yellowing of leaves called chlorosis

- Phosphorus: It helps plants hold and transfer energy for metabolism and also affects cell division, root development, and flowering. Deficiency symptom includes purple coloring of leaves or stems

- Potassium: It activates enzymes and regulates opening and closing of stomata. It also regulates water uptake by root cells

- Calcium: It regulates transport of other nutrients into the plant and is involved in the activation of certain plant enzymes

- Magnesium: It is important part of chlorophyll and is important in the production of ATP through its role as an enzyme cofactor

- Sulphur : It is a structural component of some amino acids and vitamins and is essential in the manufacturing of chloroplasts

- Silicon: Silicon is deposited in cell walls and contributes to its mechanical properties including rigidity and elasticity

- Iron: It is necessary for photosynthesis ; Present as an enzyme cofactor in plants

- Molybdenum: It is a cofactor to enzymes important in building amino acids

- Boron: It is important for binding of pectins in the RGII region of the primary cell wall and also it plays a significant role in sugar transport, cell division, and synthesizing certain enzymes

- Copper: It is important for photosynthesis; Involved in many enzyme processes; Necessary for proper photosynthesis; Involved in grain production; Involved

in the manufacture of lignin (cell walls)
- Manganese: It is necessary for building the chloroplasts
- Zinc: It is required in a large number of enzymes and plays an essential role in DNA transcription
- Chlorine: It is necessary for osmosis and ionic balance; Plays a role in photosynthesis
- Nickel: It is essential for activation of urease, an enzyme involved with nitrogen metabolism and it can substitute for Zinc and Iron as a cofactor in some enzymes
- Sodium: It is involved in the regeneration of phosphoenolpyruvate in CAM and C4 plants and it can also substitute for potassium in some circumstances
- Cobalt: It is essential for legumes for nitrogen fixation and it can substitute for molybdenum

We all know that under-nourishment of plants may lead to poor growth and development. But over-nourishment is also not good for plants. Over-fertilization or adding extra fertilizer doses to plants may lead to nutrient toxicity. Five types of nutrient toxicity are Chlorosis (yellowing of plant tissue caused by a shortage of chlorophyll synthesis), necrosis (death of plant tissues), accumulation of Anthocyanins (production of a purple or reddish colorization of foliage and/or stems), lack of new growth (stunting or reduced growth) and stunted new growth. So the thumb rule in fertilizer application is as follows:

Thumb rule in fertilizer application: Thumb rule is to apply right quantities of fertilizers at the right time using a right method. That is,

1. Apply right quantities of fertilizers
2. Apply fertilizer when the plants can best use the nutrients
3. Apply small amounts of fertilizer frequently
4. Be careful not to over fertilize

Integrated Pest Disease Management: Another important organic growing practice is the use of IPM for pest management. Integrated Pest Management or IPM is a holistic approach for the control of pests using all available pest control practices such as cultural control, mechanical control, biological control and chemical control of pests while ensuring the safety of the foods and the environment. Similarly, for effective disease management, we can use IDM (Integrated Disease Management) and for effective weed control, we can use IWM or Integrated Weed Management.

Isolation Distance/Barriers: An important practice in organic farming is the maintenance of a barrier or boundary around the organic farm/garden in order to prevent the infestations from surrounding conventional fields/gardens.

Figure 34: A Farm Fence

Beneficial Cultural Practices for Organic Crops: A number of eco-friendly organic growing practices such as trap crop technology, crop rotation, intercropping and companion planting, use of biocontrol agents and biopesticides, and growing green manure crops *in situ* may be adopted by organic growers for the success of their organic farming ventures.

Aphids and similar insects can be effectively controlled by introducing their natural enemies/predators such as ladybug beetles.

Figure 35: Lady Bug Beetle

Trap crop technology uses introduction of trap crops such as marigolds and neem trees in the growing fields. These plants are believed to have insect-repellent properties.

Figure 36: Marigolds

HYDROPONICS OF GINGER AND TURMERIC

All members of ginger family are suitable for growing in hydroponics growing systems. Hydroponics is a practice of growing plants in a nutrient-rich solution, without soil.

Today the term 'hydroponics' has become synonymous with 'soil-less production of crop plants' though the term itself means that it is water ('hydro' means 'water') that is at work ('ponics' means 'labor' or 'work'). The term 'hydroponics' may seem to represent a sophisticated process but in reality, hydroponics is a simple process of crop production. The only difference of a hydroponics crop production from that of a traditional method is that hydroponics makes use of a nutrient solution as plant growing medium instead of soil.

Hydroponics under controlled environmental conditions (i.e. greenhouse hydroponics) is more successful than that of outdoor hydroponics. When higher yields per unit area and higher productivity per plant are desired, greenhouse hydroponics is preferred than outdoor hydroponics. In fact, for growing herbaceous plants like ginger and turmeric, greenhouse hydroponics is proved to be more successful than outdoor hydroponics.

Figure 37: A Garden Greenhouse

Hydroponics is based on the principle that plant growth in a traditional soil-based production system is not dependent on the soil rather it is dependent on the nutrients and moisture present in the soil. So, if the plant nutrients and moisture required for the plant growth are provided through any other medium other than the soil, plants can still have a natural growth. Therefore in a hydroponics system, ideal nutrient and moisture requirements of the plants are fulfilled through a water culture or solution culture under ideal environmental conditions. In other words, hydroponics is *soil-less crop production system under controlled environmental conditions*.

There are mainly two types of hydroponic grow systems: Aggregate systems and DWC (Deep Water Culture) systems. Aggregate systems use different types of aggregate growing media as plant root support system while DWC systems use only water culture or nutrient solution to grow the plants. These systems do not use any aggregate growing media as plant root support system.

In DWC systems, the plant roots are totally suspended in the nutrient solution. A tray made of plastic or Styrofoam boards or similar materials that float on the surface of the solution is used to support the plant above the solution. Holes are provided on the tray so that the roots are inserted into the solution while the shoots stand on the tray growing upwards. In DWC systems, the nutrient solution needs to be aerated or bubbled continuously by using an air pump and air stones. Nutrient solution should be changed regularly and kept at constant level in the reservoir tank. Best examples of popular DWC systems are Nutrient Film Technology (NFT) and BubblePonics. BubblePonics or bubble hydroponics is water culture hydroponics where constant oxygenation or aeration of the plant root zones is required for the healthy growth of the plants.

A standard DWC hydroponics grow system is divided into TWO areas: a propagation area (propagation room) and a grow area (grow room). Propagation area is where propagation takes place while grow area is exclusively dedicated for vegetative growing purposes.

A standard hydroponics system includes grow trays, reservoir tanks, plant pots, plant nutrient kits , pH test kit, digital TDS meter, water pumps, air pumps and air stones, air diffusers, growing media kits, starter materials (seeds, clones etc), and a plant starter tray. The system should have air filters, duct mufflers or silencers, and a light proof system .Safety of operations is ensured in a hi-tech hydroponics system by providing provisions for fire protection, and insulation.

It is advised to start planting materials in the hydroponics system itself. This helps avoid transplant stress for the plants. Propagation within the hydroponics system will also help the growers obtain disease-free and pest-free starting materials. A separate propagation room may be used for propagation purposes.

For commercial hydroponics of ginger and turmeric, DWC system of hydroponics may be the best choice. A controlled environment such as a greenhouse is recommended for growing them because such a system provides shelter, and stress-free environment for the plants. Temperature of the growing environment may be monitored regularly by using a thermometer. Temperature within the growing environment should be kept at 70- 85 degrees Fahrenheit.

A hygrometer may be used to measure the humidity inside the growing environment. Humidity may be kept at 70-90%. A well-designed hydroponics system has cross air flow system to ensure adequate aeration around the plants and their root zones.

Preparation of Nutrient Solution: Preparing nutrient solution for plant nutrition is an important step in hydroponics. For healthy plant growth, a plant needs both macronutrients and micronutrients (trace elements). Major macronutrients include Nitrogen (N), Phosphorous (P), Potassium (K), Calcium (Ca), Magnesium (Mg), and Sulphur(S). Trace elements are Iron (Fe), Manganese (Mn), Boron (Bo), Zinc (Zn), Copper (Cu), and Molybdenum (Mb).

Nutrient formulas containing all these nutrients in correct proportions are available in the market as hydroponic nutrient mixes. A grower may purchase them to prepare the nutrient solution for the hydroponics.

Regarding, preparation of the nutrient solution, care should be taken that only good quality water is used. Hard water should be avoided by all means. Periodical flushing of the nutrient solution is necessary to prevent salt build up in the solution.

While preparing an ideal nutrient solution, pH, electrical conductivity (EC), temperature and total dissolved solids (TDS) of the solution should be measured as each of these parameters has an impact on the degree of nutrient absorption by the plant roots.

Optimum pH range for the nutrient solution should be kept at 5.5-5.8. Large variations in the nutrient pH may lead to poor absorption of nutrients by the plants. pH of the nutrient solution can have a great impact on the plant growth. Since every plant has a preferred pH range at which plant nutrients become available to its growth, solutions having too low or too high pH should be avoided in a hydroponics system. pH of nutrient solution should be checked regularly by using any of the pH devices available in the market. Low cost pH devices such as a pH control kit and pH pen may serve this purpose for those who are looking for cost effectiveness. A pH meter may be a costly device as compared to a pH control kit but provides instant reading.

Electrical Conductivity (EC) of nutrient solution should be between 2.2 and 2.6. EC refers to 'electrical conductivity' or flow of electric current through the nutrient solution. EC and concentration of the nutrient solution is proportionately correlated. i.e. when the concentration of nutrients is higher in the solution, EC will be higher and vice versa. EC meter is used to measure the electrical conductivity of the nutrient solution. EC meter records the reading in either micromhs per centimeter (uMho/cm) or microsiemens per centimeter (uS/cm).

The temperature of the nutrient solution affects the reading of the EC meter. Hence it is recommended that EC should be measured at 25^0 C always. If the temperature of the nutrient solution is above 25^0 C, the EC reading will be higher, even though concentration of the solution remains same. If the temperature of the nutrient solution is below 25^0 C, EC reading will be on the lower side.

Figure 38: Ginger Flower

Total Dissolved Solids (TDS): TDS refers to the total dissolved solids present in the nutrient solution. A TDS meter is used to measure TDS level of the nutrient solution. The meter reading is shown in parts per million (ppm). A detailed account of nutrients present in the nutrient solution is as given below:

1. Nitrogen: 170-180 ppm
2. Phosphorus: 110-120 ppm
3. Potassium and Calcium: 200-240 ppm each
4. Magnesium 40-55 ppm
5. Iron: 4-6 ppm
6. Manganese: 3 ppm
7. Zinc: 0.25 ppm
8. Boron: 0.70 ppm
9. Copper and Molybdenum: 0.06-0.07 ppm each

In hydroponics, the roots of the growing plants should be aerated at regular intervals because plant roots need oxygen in order to survive. This oxygenation of roots may be carried out by using highly efficient air pumps and air stones. *Remember, good aeration of the hydroponic solution is essential to obtain the best results.* If nutrient solution is poorly aerated, it adversely affects the root development. A healthy root system is white coloured and highly branched. In poorly aerated solution, roots develop browning and turn to dark colour. Regular monitoring of the nutrient solution provides the grower an idea about the status of the nutrient solution.

Regular testing of the nutrient solution for pH, EC, and TDS helps growers ensure that plants are being fed with right nutrients at right concentration. It also helps monitor the salt levels of the nutrient solution at every phase of crop production. Thus growers can take appropriate corrective measures if the salt levels rise unexpectedly resulting in a 'salt build-up' in the growing system. Two important corrective measures recommended for eliminating problems associated with salt build-up is regular flushing of the growing medium with a fresh nutrient solution or replacing the nutrient solution with a fresh one. *Remember, temperature of the nutrient solution should always be lower than the air temperature.* A few tips for successful nutrient application is given below:

1. Whenever a nutrient solution is prepared, use a measuring cup to take correct quantities of nutrients
2. Make sure that nutrients are taken in correct proportions
3. If nutrients are in powder form, use warm water to dissolve them and mix well by stirring vigorously to get a homogenous solution
4. Use a PPM (parts per million) measuring device (e.g.:

nutrient monitor) to measure the concentration level of the nutrient solution

There should be an air filtration system in the growing system. Carbon filters effectively eliminates undesirable odours from the grow system. Adequate lighting should also be provided for the growing plants@16 hours plus of light/day. Fluorescent grow lights or LED lights may be used for providing artificial light. LED lights are highly energy efficient and economical. LEDs are low temperature way to increase the amount of light that plants receive. Light distribution and coverage within the system can be adjusted by installing panels and reflectors. Placement of the lights should be directly related to the intensity of the light required by the plant. If more light intensity is required, place the light close to the plants but not too close to burn the leaves. Adjustable lighting system may be used to adjust the light according to the plant requirement. In hydroponics, care should be taken not to expose the roots of the growing plants to the light. Root exposure to light may induce growth of algae and thus contaminate the growing medium. CO_2 forms an integral part of a plant growth system and therefore it is important that CO_2 should be applied in a hydroponics system for healthy plant growth. Generally CO_2 is administered to the plants through a tank application process.

Significance of Light Energy and CO_2 in Hydroponics: Plants need light energy for various purposes, major being photosynthesis and transpiration. During photosynthesis, plants produce carbohydrates (foods) using light energy, carbon dioxide and water. In an enclosed hydroponics system, artificial lighting system and CO_2 application system may be used to provide the light and CO_2 needed by the plants.

In indoor hydroponics in greenhouses, pests and disease can also be effectively controlled by protecting the greenhouses from the entry of insects and pests. Wire meshing may be used at all entry points to prevent pest infestation. If proper hygiene is practiced within the greenhouses, disease incidences can be minimised to nil. In well-managed hydroponic grow systems, a grower can produce a good quality crop of ginger in 8-9 months. Both quality and quantity of the crop can be optimised in a well-managed hydroponic grow systems.

In fact, various types of hydroponic grow systems for ginger and turmeric are still being tested by researchers around the world. Some of the researches in this field reveal that a yield of 3-6 lbs of ginger rhizomes/square foot is possible in greenhouse hydroponics after 9-10 months of growing, provided that a plant spacing of 1 foot between plants and 1.5 feet between rows is followed.

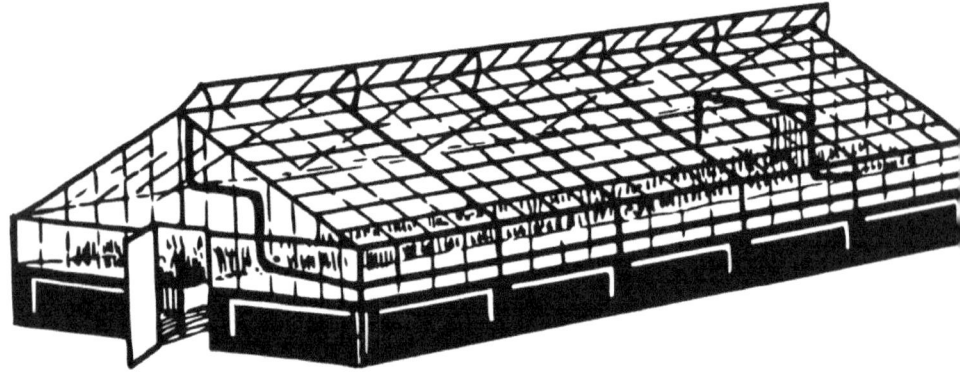

Figure 39: Structure of a Greenhouse

Bibliography

Romand-Monnier, F. & Contu, S. 2013. Curcuma caulina. The IUCN Red List of Threatened Species. Version 2014.3.

Retrieved January Firday, 2015 from http://www.kew.org/science-conservation/plants-fungi/curcuma-caulina-indian-arrowroot

USDA Nutrient Database. (2015, January Firday). Retrieved January Firday, 2015, from USDA Nutrient Database: http://ndb.nal.usda.gov/ndb/search/list

ABOUT THE AUTHOR

Roby Jose Ciju is the author of *'The Art of Perfect Living: The 7 Personal Powers for Perfection'*, an inspirational book based on scriptural wisdom. She is also a professional horticulturist and an agribusiness consultant with a Masters Degree in Horticulture and a Post Graduate Diploma in Agri-Supply Chain Management.

She has founded https://agrihortico.com, a website dedicated for publishing information on Food & Agriculture Topics. She has written more than 40 books on various Food & Agriculture topics till date and her best-selling books are, Mushroom Farming, Moringa, Curryleaf, Jalapeno Peppers, and Growing Ginger, Turmeric and Arrowroot. She may be contacted at roby@agrihortico.com. You may follow agrihortico at https://twitter.com/agrihortico1.

Roby Jose Ciju

www.ingramcontent.com/pod-product-compliance
Lightning Source LLC
Chambersburg PA
CBHW070828180526
45168CB00002B/773